AN ADDISONIAN PRESS BOOK

Library of Congress Cataloging in Publication Data
Niizaka, Kazuo
Clouds

"An Addisonian Press book."
SUMMARY: Relates in verse the variety of cloud formations two children spy one afternoon while watching clouds in the sky.
[1. Clouds—Fiction. 2. Stories in rhyme]
I. Stanton, Henry B. II. Title
PZ8.3.N566Cl [E] 74-19155
ISBN 0-201-00574-3

First published in the United States 1975, by Addison-Wesley.
Printed in Japan

First Printing

First published by Shiko-sha Company Ltd., Tokyo, Japan

Clouds

BY KAZUO NIIZAKA

TEXT ADAPTED BY HENRY STANTON

ADDISON-WESLEY

Mary and Tom and Puff-the-dog
Go for a gallop one afternoon—

Puff, Mary and Tom
(and Tom's balloon).

Mary and Tom grow tired at last.
Down they flop in the cool, green grass—

and watch the fluffy clouds drift past.

But Mary's dog Puff and Tom's balloon
Aren't tired at all—it's scarcely noon.

Puff scutters away to chase a shoo-fly.
The balloon floats off, up into the blue sky.

Mary sees clouds that look like moons,
like fat raccoons,
like white balloons.

Tom sees clouds that look like snowballs,
just like soft, white I-can-throw-balls.

Isn't it odd and isn't it funny.
You look at a cloud and you see—a bunny

. . . or a cow with horns
. . . or a pair of mittens
. . . or a big white cat with little white kittens.

But Tom and Mary rub their eyes.
For the cloud *they* see is a real surprise.

Puff-the-cloud, like Puff-the-dog,
Comes trotting through the sky.
His wooly tail and wooly ears
And woofly nose held high.

And then come children
Playing tag.

See how Puff's tail
Starts to wag.

Oh, mercy me! A crocodile!
With gaping jaws
With toothy smile.

Tom thinks this is really scary!
And—
"Look out, yellow bird!" cries Mary.

And now we have a monster whale.
Its tail goes swish-swash-swish.

It is laughing at Tom's red balloon
Swimming beneath it like a fish.

An elephant, a lion—a zoo of clouds—
Come drifting through the blue, blue sky.
Following the red balloon
As it goes bob-bob-bobbing by.

And up above, the yellow bird
Sees the wind-blown race.

Will the clouds
Or will the bird
Win this merry chase?

Pop!

Bang!

Oh bungling bird!
Oh burst balloon!
Oh see the feathers fly!
And see the
Crazy mixed-up clouds
Go tumbling through the sky!

The game is over. Off they go,
The whole gallumphing crowd!

All except for Mary's friend,
poor old Puff-the-cloud.

Puff has a surprise for Mary and Tom.
Where do you suppose he got it from?

And what do you suppose it is—a rag?
A flower, a shoe, a paper bag?

A skipping rope? A bunch of cherries?
A piece of soap? A hat of Mary's?

"Well, anyway, anyhow," Mary said
"I think it's sort of small and red—
        I think it's a piece of your balloon."

Mary and Tom and Puff-the-dog
Went for a gallop one afternoon—
Puff, Mary and Tom
(and Tom's balloon).

Though balloons may burst and clouds don't last
It IS fun, watching the clouds drift past!